poetry is a life-cherishing force...

—mary oliver

i think there's a kind of desperate hope built into poetry …
one really wants, hopelessly, to save the world. one is trying
to say everything that can be said for the things that
one loves while there's still time …

—w. s. merwin

EARTH TO POETRY

a 30-days, 30-poems
earth, self & other care challenge

l.l. barkat

masters in fine living series

ts T. S. Poetry Press • New York

T. S. Poetry Press
Ossining, New York
Tspoetry.com
© 2019, L.L. Barkat

Cover image by Sonia Joie.
soniajoie.com

ISBN 978-1-943120-33-8

Library of Congress Cataloging-in-Publication Data:
Barkat, L.L.
        [Nonfiction. Poetry. Writing. Climate. Self Care.]
        Earth to Poetry: A 30-Days, 30-Poems Earth,
            Self, and Other Care Challenge/L.L Barkat
        ISBN 978-1-943120-33-8
        Library of Congress Control Number: 2019911186

for those with the will,
i wish you the way

&

for my daughters, who care
worlds upon worlds for this earth,
because it's our beautiful home
and this is where we
love each other

table of contents

a way in

At its core, caring for the earth is a simple matter of caring for one's natural home, caring for oneself, and caring for others we love. It's easy to forget this, in the face of what can sometimes feel like the complex nature of earth-care solutions, as well as varying philosophical perspectives.

But, really—to care, or not to care....that is not the question. After all, as humans most of us care quite deeply about our surroundings, ourselves, and our loved ones. We just don't always sense how widely our lifestyles and societal design can affect the earth, which includes, in effect, ourselves and others. Or we aren't sure where to begin.

This 30-days, 30-poems challenge assumes we care—or want to—and it helps us find *a poetic way in,* as well as understand our hopes and struggles in a world where sometimes we've wondered: does anyone care? They do. And they can.

Caring takes imagination; the kindling of our hearts; and an ability to see and cultivate connections, both obvious and surprising. From there, it's a matter of finding interesting and exciting solutions that feel life-enriching; even—shall we say it?—sometimes *fun.*

Reading and writing poetry provides a start. So, let's read, and write, and imagine, and *live.* Plus, whenever we can, let's have a bit of fun.

1

day one: begin

Poem Reading

Find a single
tree, find
the moon.
It doesn't
take much.
Just begin.

—L.L. Barkat, from *God in the Yard*

Prompt: Begin

Focusing on a single natural item or two was (and still is) the simple basis for a longstanding genre of poetry: haiku.

Imagine that! A whole genre that relies on attentiveness to our natural surroundings.

The haiku understands that it doesn't "take much" to start the mind going and get the writer to start writing.

It may take a little more for us to craft something truly beautiful, insightful, even startling. But that doesn't matter at first. The first thing is to begin.

Today, write a poem that tells its reader to find a single

thing located outdoors, as a starting point for reflection, love, or inquiry.

You can begin your poem like the one in the reading above, if you wish: "Find a single..."

In the spirit of haiku, stay with just one or two images. Really attend to them. Write short. Then, if you want, write long. How does the medium of *poem length*—the poem's design—change the poem's message or your experience of it, if at all?

2

day two: a dawn

Poem Reading

After Reading About the Iceberg Waterfalls

> *Today I am startled again because it now appears that*
> *the ocean will likely be free of summer ice by 2040…*

—*Jennifer A. Francis, Scientific American*

Melting,

my heart is.
Falling.

Careening

over the edge
of what is

and what will
be.

Come with me,
friend.

The sea

is rising,
in a daughter's dawn
of time,

to meet us.

—L. L. Barkat

(*Best case scenarios place the complete disappearance of summer Arctic ice at just 21 years out. My daughter is about to be 22.*)

Prompt: A _____ Dawn of Time

In the above poem, I've measured time by the age of my eldest daughter, noting that Arctic summer ice will be gone, causing the sea to rise significantly, "in a daughter's dawn of time"—about 20 years from now.

Write a poem in which you measure time in terms of a family member's, a friend's, or a neighbor's age.

Or, write a poem in which you promise a family member, neighbor, or friend that, because of your love for them, you will do one or two specific things to help slow the rising of the seas. (Need ideas on what those things might be? See the section on p. 82: The Top 80 Game Changers Shared in *Drawdown*.)

3

day three: just when

Poem Reading

The earth's economy

Just when I thought the day
had nothing left to give,
when heat was ladled across
the shallow dry plate

of the nation, working or not, alive
or not, my country
road home from work
an affair of sour radio news and roadkill —

the furred skunk, possum, cat,
squirrel, raccoon, in the
special economy of the outward-
facing nose, lost in final scent,

the surrendered open mouth,
forehead pressed back in frozen
tragedy, tension gone, time done,
appetite dissolving into skull —

I find myself at the kitchen counter
in a different Americana, tearing
kale ruffles from their spines
for a chilled supper of greens with lemon

and oil, Dijon, garlic, cucumber —
live, wet and impossibly cool from the
earth garden just outside the door,
where the farmer's wife one hundred

years ago also opened her apron
like a cradle, gingerly receiving
into thin billowing cotton pockets
as much as she could carry

as much as she could carry

—Ruth Mowry, from *The Joy of Poetry*

Prompt: Just When I Thought the Day

Writer Lorraine Hansberry has said, "There is always something left to love."

Similarly, there is always something left for which we can feel gratitude—something to love about our days, despite our occasional brushes with doubt or despair.

In "The earth's economy," the poet begins, "Just when I thought the day/had nothing left to give." She goes on to name experiences and realities that have made her feel empty and despairing. But then the day gives her something to feel gratitude for, and it is no little gratitude. The gift is abundant, and she is free to take as much as she "could carry."

Write a poem that starts, "Just when I thought the day had nothing left to give." Name your doubts or despair. Then name something you can feel gratitude for, some gift from the earth's abundant economy.

4

day four: age

Poem Reading

Back

I don't know how to age
a tree, without cutting it open
to count the rings.

This tree in front of the tan house,
taller than its two stories,
is it older than me?

Did it once watch my yellow hair,
my skinny legs grow,
and cover me while I drank

Kool Aid in July
and tried to decide
whether to follow

my brother
or my sister
around that day…

—Will Willingham

Prompt: *Age a Tree*

In "Back," the poet says, "I don't know how to age/a tree, without cutting it open…" The rest of the poem then becomes an exercise in aging a tree—not by cutting it down to count its rings, but rather through memory.

Write a poem that ages a tree through memory. It could be a tree from your childhood. Or it could be a tree that's been part of your adulthood. Consider looking up details about your chosen tree—its scientific and common names, where it grows best, its history, etc.

5

day five: prayer or ode

Poem Reading

Prayer for the Pompadour Looking Up From My Plate

God, I feel bad.
He died for me
and I could only
eat half of him.

—Marcus Goodyear, from *Barbies at Communion*

Prompt: Prayer for or Ode to Food

Write a prayer for or an ode to a food item you are about to eat or already ate today. Will you (or did you) eat the whole item? Why or why not? If you aren't going to eat the whole item (or didn't), what will happen to the item once it's no longer on your plate? You can explore the answers to these questions in your poem, too.

6

day six: no rain

Poem Reading

Aubade

I still want you,

though it hasn't rained in forever,
and we have lost our sun umbrella.

The road is closed, the one
that leads to the sea,

where no one cares about the rain,
nor the lost umbrella,
nor these words that whisper,

wanting.

—L.L. Barkat, from *The Novelist*

Prompt: No Rain & Road

Write a love or friendship poem that captures a sense of loss. Include both:

- some kind of natural element
- some kind of human-made element

For example, "Aubade" uses the images of *no-rain* and a *closed road*.

7

day seven: distances

Poem Reading

North on the Illahee Ferry

So there you are, where you wanted to be.
I can imagine the Seattle city pier falling away
behind you. Herring gulls wheel along their wires,
reflections shattering in the ferry wake. You lean out

over the swell, caught by blue distance, and when
the cold finds its way onto the deck, plunder a pocket
for an orange and break the body into crescent-shaped
pieces brought in a wooden cage over the pass

from their God-hung green night. Teeth tear membrane
as the coastline recedes. North is sudden and near,
the final island before dark. Urgently the senses surge
toward Texas, parts south where an orchard yields

the finite sun over and over again, where once
someone loved you. Swallow what reminds you of home.
What's held in both hands and the limitless motion
you longed for and only dimly understand—love,

the same thing sustains them, the vastness
that's kept you and indeed everyone on the vessel afloat.
And at this table where a letter has reached us,
is being read and reread, we are nearly present on the boat,
when both citrus and the salt season the moment.

—Anne M. Doe Overstreet, from *Delicate Machinery
Suspended*

Prompt: Distances & Magnitude

Images of both distance and closeness, smallness and vastness, exist in "North on the Illahee Ferry." Every life is sus-
tained by both.

Write a poem about an experience that is very close or
small, the way the ferry, the pocket, and the orange are to the
ferry rider in the poem. Every few lines, pull back and consider the distance or vastness that supports the possibility of
this close or small experience.

Such a poem might be the perfect candidate for the villanelle
form, with its built-in repetition mechanisms that can help you
travel between two realities such as closeness and distance or
smallness and vastness.

Instructions for a villanelle can be found at:
https://www.tweetspeakpoetry.com/2011/12/02/i-see-you-
in-there-the-villanelle/)

8

day eight: pretend

Poem Reading

Today by the Creek I Pretend I'm the Dalai Lama

Today by the creek I pretend I'm the Dalai Lama.
I don't know enough to get it wrong.
The robes saffron, golden.
The smile, serene. The walk
feline, Persian, with a hint of bobcat.
My breath comes from the belly,
sanctifies plants that sway on the out-breath,
spring seeds released by my presence.
My imagined sandals lined in black
cotton tread lightly. People smile,
they namaste and bow with tears in their eyes.
As Dalai Lama I eat lightly. I forego coffee,
share my morning oatmeal, split my chocolate
enviro-mint with Rose at Chico Natural Foods.
I bestow smile after smile, hands
raised in blessing as the sun blesses
my eyelids, the frog blesses my ears, a twang
on the base fiddle, a strum on the steel guitar.
Small beetles spin as they mate,
tail ends merged in bug ecstasy.

And as Lama, celibate as a leaf, myself,
I smile at them.
One water strider jots back and forth nearby
wondering, no doubt, is it he? Is it really he?
Our neutral, kind, enlightened Dalai Lama?
The Kingfisher asks, and other birds,
I can hear it in their calls—Is it he? Is it he? Is it she?
Is it she? She she she she chi chi chi chi chi…
I bow my head in gratitude
toward the pill bug in my leg's shadow.
I love it unconditionally.
I love being Dalai Lama.
I can't wait for night with Dalai Lama dreams.
On my knee one of those firefly-looking beetles
with orange eyes and black wings stretches
in a bug yoga of oblation, antennae a-twirl,
wings open in sun salutation in honor of my presence.
And as Dalai Lama I am one with it.
For today, just today, as Dalai Lama, I declare it.
In spite of everything, villages in flames
murders, walls and wars,
I declare it, I declare it. I am Dalai Lama, today,
here by the creek,
and all is well.

— Susan Wooldridge, author of *poemcrazy:*
    freeing your life with words

Prompt: I Pretend

Think of the kindest public or religious figure you know, either in present times or history. Consider their aspect: how they dress (dressed), their looks, their way of walking or other gestures. Pretend you are this person, walking through both the natural and the human-constructed world.

What details do you notice as you walk? Be specific, as Goldsmith has been specific in her poem. How do you extend kindness along the way? How do you express gratitude? How do you acknowledge, with specific acts, the creatures and environs you pass through? How do they respond to you?

9

day nine: price

Poem Reading

The Poetry of Money 2

for Rachel Marks

The tail of the peacock male
shimmers with metallic coins,
the eyes of money winking
beauty's price. What is a poem
worth? Just this feathery spread
melting change into cool rain.

—Marcus Goodyear, from *Barbies at Communion*

Prompt: Beauty's Price

Using natural images, write a poem about beauty's price. How
does beauty encourage us to take care of it, thus asking us to
pay a cost for keeping it intact? Is it worth the price? Or, how
does beauty take care of us, thus making beauty itself pay the
cost of care?

Alternately, write a poem about what a poem is worth, using natural images to express that worth. ("The Poetry of Money 2" includes a peacock, feathers, and cool rain.)

10

day ten: look

Poem Reading

Drought on the Open Road

Once the herd was so thirsty
they ate the burn right off
the interstate shoulder, two bites
from asphalt and cars flying
75 miles to nowhere.
Heat paralyzed cows
never look up.

—Marcus Goodyear, from *Barbies at Communion*

Prompt: Look Up

Write a poem about ordinary needs that sometimes lead us to
do or consume what is not natural or good for either us or
our natural surroundings. Why don't people "look up"? In
other words, why do they almost blindly do things that are
counter to their well-being or survival?

Also, for an extra challenge, play around with line breaks,
as well as line lengths, to create either…

• a sense of doing something mindlessly, or

• a sense of creating sudden attention that could cause someone to "look up" and reflect

11

day eleven: love

Poem Reading

Awaiting Arrival

After all these years of nothing coming
down the road, she still preserves lemons,
still salts fish and leaves the door unlocked
to the winter traveler. In the pantry
among the Ryvita rye and sesame crackers,
there are rooted things in abundance:
the fat pearl of Walla Walla sweet onions,
the folded wings of garlic, each shrouded
in tissue paper. She, like all her neighbors,
keeps the freezer stocked with zucchini bread
and coffee in case of the improbable arrival
of snow or the advent of a stranger, each visitation
anticipated like a bridegroom a long time coming.

—Anne M. Doe Overstreet,
  from *Delicate Machinery Suspended*

Prompt: Food as Love

Food is the center of hospitality in "Awaiting Arrival," even though there's no particular guest expected on the horizon. Each improbable visit is anticipated even "after all these years of nothing coming down the road." Like many of us, the people in the poem long for connection.

Write a poem in which you describe the foods you or a friend or relative keep on hand, in case of someone's arrival. Are they foods that have been "put up" as a result of our earth's abundance (like the zucchini bread in the poem)? Or are they foods that have been packaged and provided by vendors? Do you (or a friend or relative) tend to over-prepare or under-prepare for the possibility of company? (In a stark example of minimalist preparation, check out Anna Akhmatova's "carrot and plate offering" at tweetspeakpoetry.com/2018/12/06/literary-friends-keeping-anna-akhmatova-alive/.)

If you like, begin or end your poem with the words, "After all these years of..."

12

day twelve: influences

Poem Reading

Father's Day

Remember the tree house?
I suppose that was less us
than perhaps the music
at church, or the car—my bad,
bad cars. But remember anyway.

There were the grass fires
we saw when I was young.
You don't know this yet, but
I've written about them,
the smell of smoke and vanilla.

The business trips you took
us on. The short stories said
while waiting at the post office.
How you tried to convince us
that camping was fun; it was

in retrospect. The tree house,
hung from an ageless pine,

provided a new perspective
on everything I saw from ground-
level, our whole backyard.

—David K. Wheeler, from *Contingency Plans*

Prompt: Influences

Our stance towards our surroundings has been, of course,
influenced by our parents and siblings or friends.

How do you see our natural world and your place within
it? Do you tend to spin "grass fires" into "the smell of smoke
and vanilla"? Or do you tend to need convincing that "camping [is] fun"?

Have you had some kind of "tree house, hung from an
ageless pine" that allowed you to live fairly seamlessly with
your natural surroundings?

Write a poem about a parent, sibling, or friend who has influenced your stance towards our natural world. You can begin
your poem with a question, as in "Father's Day": *Remember
the…?*

13

day thirteen: question

Poem Reading

Petit à Petit L'Oiseau Fait Son Nid

Little by little, they say,
the bird makes its nest.
I have been making mine
in silvered hemlocks, time
after time; today I used a red
thread I found near the garden.

I used to dream of living in a garden,
listening to words white orchids say
to emerald hummingbirds, red-
throated, stealing gold for nests
the size of women's thimbles, time
beating between breaths, a rhythm mine

could never find trapped, as in a mine
long hollowed, tapped black garden
that metamorphosed over time,
caught sounds of earth-on-earth say,
Come bed yourself on rock-hard nest,
turn death to sapphire, diamond, ruby red.

Rumor spreads: inside the earth is red,
molten, thrusting gold like mine
into the sun, into evening's nest
that sits above an empty garden
where orchids do not say
it is time

it is time
to ravel rays from ravished dreams, red
and unremembered; it is time to say
what is yours and what is mine
it is time to turn the garden
into earth, find fool's gold for a nest.

I have been making such a nest,
little by little, time after time,
I have been dreaming near a garden
in threads of memories, ruby red.
I have been claiming what is mine
and inviting you to say

you want the nest, the gold turning red,
the time we knew was mine,
the garden waiting, for what you have to say.

—L.L. Barkat, from *Love, Etc.*

Prompt: Working Out a Question

Even if you've never written a sestina before, today might be the day to give it a try. Because of the particular way sestinas repeat end-of-line words, they're a great form for dealing with something you're trying to work out in your mind—a question or a problem.

The "Petit" poem uses a garden setting complete with white orchids, emerald humming birds, and silvered hemlocks to work out a life question.

As you work out a question or problem in your poem, include a natural setting: the seashore, woods, a secret place in a city park, the desert, mountains, or something else. Include some specific details that will make the setting come alive. Do a little research if you need to. It's easiest to research by region, when looking for specifics. See links below, for instance.

Instructions for a sestina can be found at:
https://www.tweetspeakpoetry.com/2011/06/27/write-your-first-sestina-its-a-matter-of-pride/

Research by U.S. region:

11 Great U.S. National Forests
https://www.wilderness.org/articles/article/11-americas-greatest-national-forests

Central Park, NYC
https://en.wikipedia.org/wiki/Central_Park

List of Mountains in the U.S.
https://en.wikipedia.org/wiki/List_of_mountains_
of_the_United_States

New Fields Indiana (Museum of Art)
https://discovernewfields.org/

Puget Sound
https://en.wikipedia.org/wiki/Puget_Sound

The Big 4 North American Deserts
https://rootsrated.com/stories/a-captivating-
look-at-the-big-four-north-american-deserts

14

day fourteen: technology

Poem Reading

Eve's Second Garden

Filled with succulents and spines
that scratch her legs and break
off inside, so Adam has to pinch
her skin, then suck and bite
the sliver, his teeth the only tweezers
yet invented. Nothing requires much
water. Five months drought each winter
turns prickly pears brilliant purple
before they die. Yucca shade hides
the snakes, so she rustles the grass
with a sotol stalk. Frightened, they rattle.
She freezes and listens to them promise,
"Our kisses turn you brilliant purple."

—Marcus Goodyear, from *Barbies at Communion*

Prompt: Guardian Technology

In today's poem, we see a couple using simple technology (teeth, a sotol stalk) to save and protect themselves from variables in their environment that are, rightly so, frightening.

Write a poem about something in our environment that frightens people (or frightens you). Include some kind of technology they (or you) have used to save or protect themselves. How does this technology compare to the type of technology in the poem?

If you like, make a comparison of technology types, or ask a question that helps us reflect on our technological responses to fear.

15

day fifteen: sun

Poem Reading

Papa is the sky
Mama is the air
Elaina is the clouds
Zoya is the moon
And Polly is the sun

—Zoya Marchenko, age 5, from *Sun Shine Down*

Prompt: You Are My Sun

Write a poem about your family, a group of friends, your co-workers, or even parts of your own self and roles that assigns some kind of natural phenomenon to each.

In the poem above, five-year-old Zoya very simply says that Papa is the sky, Mama is the air, and so on.

Make your poem as simple as Zoya's, using hers as a template. Or, get more complex and put your metaphor in a form poem like the ghazal or the sonnet.

Instructions for a ghazal can be found at:
https://www.tweetspeakpoetry.com/2013/10/07/ghazal-poetry/

Instructions for a sonnet can be found at:
https://www.tweetspeakpoetry.com/2011/08/08/glass-slipper-sonnets/

16

day sixteen: deep-downness

Poem Reading

> Do the shells still hear the sea,
> though they are in pieces;
> how deep does the hearing of the sea
> enter into bone...

—L.L. Barkat, from *The Novelist*

Prompt: The Deep-downness

Today's poem considers the deep-downness of things. What is retained within something, even when it is removed from its ordinary setting or becomes broken?

Write a poem that explores some kind of deep-downness, whether it be in stardust or melting icebergs or even some aspect of yourself.

Start with a question, if you like: "Do the...?" or "Does the...?"

It doesn't matter if the answer is yes or no. What matters is that you ask. Honest questions always lead us to interesting places.

17

day seventeen: reclamation

Poem Reading

Resort

When the Baker Hotel died,
no one ordered an autopsy
or called the local mortician.
They just left the carcass
at the crossroads where it fell,
bulging brickwalls, gouged eyes,
empty sockets jagg'd with glass.
On coffee break, the local doc
doesn't wonder if he could
save her. Guilt dies without memory
so don't bother picking the bones.
The marrow sold cheap to antique
stores and left rooms mostly hollow.
Shout your name in the stairwell,
the space will keep your voice
until midnight, bubbling in sulfur baths.
whispering at windows not yet broken,
at doors stuck in jams:
Remember me. Remember me.

—Marcus Goodyear, from *Barbies at Communion*

Prompt: Reclamation

One philosophical standpoint about our environment is that we are simply part of it, and all that we build is basically like the things that all other species build; they are constructed according to our species' knowledge, our latest technologies and stylistic preferences, and our needs or wants. We even build to impress, like certain birds do.

But what about when we build in ways that don't really "fit in" with a healthy ecological scene? Often, it's at this juncture that the issue of reclamation arises.

Write a poem about an abandoned or ruined building, object, piece of land, or some other aspect of human design or living that you believe is in need of reclamation. Alternately, write a poem about a reclaiming process. (See links below for some interesting types of reclamation.)

If you like, use the end words from "Resort" to begin your poem: *remember me.*

Two examples of reclamation:

Turning Waste Water Into Drinking Water
https://www.rit.edu/showcase/index.php?id=167

Mine Reclamation
https://en.wikipedia.org/wiki/Mine_reclamation

18

day eighteen: morning

Poem Reading

Li Po

1

Li Po knew
the fecund trees
full of blossoms,
the tea bushes
flush with leaves,
sweet scent rising
from snow-petaled earth,
spears rolled–or broken–
between fingers and thumb

2

Every morning
I am Li Po,
if only I hear
the expectant cup

3

And if the porcelain overturns,
what then?

4

the snow-petaled earth
the snow-petaled earth

—L.L. Barkat, from *How to Read a Poem: Based on the Billy Collins Poem "Introduction to Poetry"*

Prompt: Every Morning

In "Li Po," the poem looks back in history to a poet who sometimes wrote about tea. It then pairs the narrator of the poem with Li Po, saying "every morning I am Li Po."

Humankind has a long history with agriculture and the harvesting of wild edibles. Choose a historical or modern-day figure involved with agriculture or the harvesting of wild edibles. In your poem, compare yourself to this figure. If you like, use the words, "Every morning I am [name of figure]..." in the poem.

Here are a few figures you might want to consider:

Steve Brill
https://www.wildmanstevebrill.com/

Jimmy Carter
https://jimmycarter.info/plan-your-visit/
jimmy-carters-boyhood-farm/

Cesar Chavez
https://ufw.org/research/history/story-cesar-chavez/

Bill Mollison
https://en.wikipedia.org/wiki/Bill_Mollison

Severine Von Tscharner Fleming
https://greenhorns.org/

George Washington
https://www.mountvernon.org/library/digitalhistory/
digital-encyclopedia/article/george-washington-and-
agriculture/

Karen Washington
http://www.riseandrootfarm.com/karen-washington

19

day nineteen: making

Poem Reading

Hawksbill Crag

By gravel road
we rise four miles
into Ozark bluff.
Our truck hugs
the slant
of timber line
thin
as a pencil streak.

At Hawksbill Crag,
we tramp
thousands of feet
above shaggy pine
and the thumb of Jehovah.

I clutch a walking stick,
while you slide to the edge
of the bluff face
and act as if you
plunge

to
your
death
until you
slip off

the boulder
back into
the buttonholes
of the pines.

—Dave Malone, from *O: Love Poems From the Ozarks*

Prompt: Making Buttonholes

Write a poem that lends human or godlike aspects and implements to things in our environment.

In today's poem reading, for instance, the poet expresses the surroundings on a hike in god- and human-related metaphors. The timber line is thin as "a pencil streak." The rocks are "the thumb of Jehovah." The pines have "buttonholes."

20

day twenty: surprise

Poem Reading

Immolation

As the horizon looms, flips over to present
an endless span of waves, I give up, surrender.
My fate's the fate of falling. I guess I hoped for recognition,
that when I pushed my arms into the hostile sun
he would look up and see my face, the frame
of limb so like his lover, perhaps invoke my name.

I imagine women fainting at the thought
of this lovely form's ravagement, the taint
of char hot enough to warp a wooden strut,
melt wax, and singe. But Daedalus flies on.
The body will soften momentarily, pliable if heavy,
finding shape hours later, so I devise my final self.

The scent surely travels downwind
in the contrail of smoke he, at least, could see.
I thought he'd catch me; if nothing else
to save the contraption with its maze

of gears and levered joints. I counted on,
I understood, he loved the thing.

—Anne M. Doe Overstreet,
from *Delicate Machinery Suspended*

Prompt: Technology's Surprise Side

In "Immolation" there's a delicate balance between constructed technology and natural forces. Each takes turns chal-
lenging the other, though it is Icarus who eventually suffers
loss.

Complex issues of the hope for recognition and adulation,
the desire for identity affirmation and love, enter the picture.
The poem seems to suggest that technology development and
technology use may not be as straightforward as we think it is,
may not always be a question of just our interaction with natural forces.

Write a poem about some kind of technology that at-
tempts to either harness or challenge natural forces, the way
Icarus and Daedalus tried to defy gravity, harness the power
of wings, and challenge the "hostile sun."

Consider: is the use of your chosen technology a straightforward matter? Or are there deeper issues related to self and
others that come into play?

You might like to try the challenge of the pantoum form,
which lends itself to tackling themes that contain layers of in-
tention and meaning.

If you like, begin your poem withe the words "I understand" or "I guess I hoped for."

Instructions for a pantoum can be found at:
https://www.tweetspeakpoetry.com/2013/03/27/
infographic-pantoum-of-the-opera/

21

day twenty-one: loveliness

Poem Reading

Night

Stars over snow,
    And in the west a planet
Swinging below a star—
    Look for a lovely thing and you will find it,
It is not far—
    It will never be far.

—Sara Teasdale

Prompt: Looking for Loveliness

In "Night," Sara Teasdale urges us to "look for a lovely thing and you will find it."

Teasdale looked and she found "stars over snow," as well as "in the west a planet/swinging below a star."

When we begin a deeper journey into earth care, sometimes we are struck by the breadth of ruin, even ugliness, that it is our challenge to recover and redeem.

While it is very necessary to acknowledge the true problems that call on our creative solutions, a continual focus on the difficulties can damage our own souls over time. Putting into place a daily or weekly practice of "looking for the lovely thing," can help sustain us and keep us creative—for, it is in a spirit of gratitude, hope and creativity that we can maintain our energy and continue to craft better and better solutions together.

Write a poem about looking for a lovely thing and finding it. You could focus on a single lovely thing and then further describe its aspects. Or, you could list several lovely things. Either way, you might want to try the catalog form, to poetically record what you looked for and found.

Instructions for a catalog poem can be found at:
https://www.tweetspeakpoetry.com/2011/05/21/how-to-write-a-catalog-poem-or-not/

22

day twenty-two: never

Poem Reading

At the end of the Sound,
where the pines have been pushed back
by an unrelenting salt wind,
you will find that jingle-shell beach—
where little cups of pearly lemon peach
stretch out endlessly. Put your hands to them
and you will not know
where to stop.

So much beauty,
so much unrelenting jingle-chiming

beauty.

—L.L. Barkat, from *The Novelist*

Prompt: Never Want to Stop

Today's poem explores natural beauty and raises the twin
issues of abundance and an unstoppable response to it.

Write a poem about some aspect of our natural, constructed, or processed surroundings that is so beautiful you never want to stop experiencing it. You can focus in on a single item if you like—a fir tree, a specific flower, a pier, a special building, tea, chocolate, silk, books, etc.

Describe the beauty in a way that features sounds you can almost feel in your mouth. Today's poem uses *j, s, ch, p, sh,* and *ing* sounds that combine to create a sense of forward motion and a desire, if one could, to have it all ("you will not know where to stop").

23

day twenty-three: signs

Poem Reading

Trouble in Paradise (excerpt)

5

Signs of loss:
the ashy fragment of a wasp's nest
bird's wing torn
roses pale
black leaves
the landscape red
the sky black.

—Maureen E. Doallas, from *Neruda's Memoirs*

Prompt: Signs Of

As humans, many of us are deeply touched by visuals and
events in the natural world, whether in times of sorrow or joy.

Write a poem in which you point to some kind of loss—
of something or someone important to you.

Alternately, write a poem in which you point to the joyous arrival of something or someone important to you.

Choose natural images—sights, sounds, tastes, textures, fragrances—that can serve as surprising and interesting signs of either loss or arrival.

If you like, title your poem "Signs of _____ ."

24

day twenty-four: favors

Poem Reading

Come, night, come, Romeo, come, thou day in night;
For thou wilt lie upon the wings of night
Whiter than new snow on a raven's back.
Come, gentle night, come, loving, black-brow'd night,
Give me my Romeo; and, when he shall die,
Take him and cut him out in little stars,
And he will make the face of heaven so fine
That all the world will be in love with night
And pay no worship to the garish sun.

—William Shakespeare

Prompt: Asking Favors

As humans, though we can sometimes feel separate from our environment, we are, ultimately, a part of it. As some say, we are made of the stuff of stars. Others note that we are water or dust. You could say that Juliet knew this and was making a special request regarding where Romeo would end up in the wideness of what makes up our environment.

In the midst of *Romeo and Juliet*, we find what amounts to a poem; in it, Juliet addresses night and asks for an eternal favor: "Give me my Romeo, and, when he shall die/take him and cut him out in little stars."

Write a poem that addresses morning, noon, or night. Ask for a favor that joins either you or a loved-one to morning, noon, or night—or even joins you or a loved-one to dust or water or some other element.

If you like, begin your poem with the words, "Come [chosen time of day or chosen element], come..."

25

day twenty-five: convinced

Poem Reading

Turning

Every rusty nail, driven into the earth
near hydrangea roots, is a trick

for turning pink petals blue.
I'm not convinced. Oh, we are

schemers and shortcut dreamers,
craving papery gills tinged

with elusive hues—periwinkle and ultramarine,
tea green, tipped with rose.

Wreaked by frost heaves,
over time, those nails erupt,

slash the bare hands of the gardener
…blood, the hot bloom of tetanus.

Is there another way?
Swamp the warming loam

with alum suspended in water—
basic salt, the most delectable seasoning.

You only get one life.

—Laurie Klein

Prompt: Convinced or Not

Earth care, as it turns out, is really about self-care and other-care. What we design today impacts how we live tomorrow. For better or for worse, it impacts far into upcoming generations.

In "Turning," a shortcut today becomes an injury tomorrow. Says the poet, "I'm not convinced," meaning that she's not convinced that every shortcut is worth the long-term effect.

Write a poem about some aspect of your living that represents a kind of shortcut that has an impact on how you or others will live tomorrow. Are you convinced about the wisdom of the shortcut? Are you not convinced? If you like, begin or end your poem with the words "I'm not convinced" or "I'm convinced."

26

day twenty-six: peace

Poem Reading

The Lake Isle of Innisfree

I will arise and go now, and go to Innisfree,
And a small cabin build there, of clay and wattles made;
Nine bean-rows will I have there, a hive for the honey-bee,
And live alone in the bee-loud glade.

And I shall have some peace there, for peace comes
    dropping slow,
Dropping from the veils of the morning to where the
    cricket sings;
There midnight's all a glimmer, and noon a purple glow,
And evening full of the linnet's wings.

I will arise and go now, for always night and day
I hear lake water lapping with low sounds by the shore;
While I stand on the roadway, or on the pavements grey,
I hear it in the deep heart's core.

—William Butler Yeats

Prompt: Peace There

Have you ever heard of the term *biophilia?* Put as simply as possible, it's our human tendency to love our natural environment and seek it out. Says Terrapin Bright Green, in "14 Patterns of Biophilic Design,"

> Biophilic design can reduce stress, improve cognitive function and creativity, improve our well-being and expedite healing; as the world population continues to urbanize, these qualities are ever more important. Given how quickly an experience of nature can elicit a restorative response, and the fact that U. S. businesses squander billions of dollars each year on lost productivity due to stress-related illnesses, design that reconnects us with nature—biophilic design—is essential for providing people opportunities to live and work in healthy places and spaces with less stress and greater overall health and well-being.

In other words, whether we are sitting in our homes, schools, or offices, or whether we are out in the park or the woods, Nature calls to something in our "deep heart's core."

We may or may not be aware of that except when our built environment really cuts us off in significant ways from our natural surroundings. Even then, our greatest awareness may simply be that we feel stressed, and we may not understand how simple the relief could be.

Write a poem that talks about the peace you feel when you go into a secret place in our natural surroundings. It should be a place where you feel relatively safe, but you can still sense some kind of mystery. If you live in a city, such a place might exist in a favorite park, a nearby courtyard, or a building that incorporated biophilic design into its aspect and landscape. Alternately, write a poem about a wide open place where you have a view for miles.

Write the peace you feel by embedding it in the images in the poem, the way Yeats embedded his peace into midnight "all a-glimmer," evening "full of the linnet's wings," and the sounds of "lake water lapping with low sounds by the shore."

For the article on biophilic patterns in design, see:
https://www.terrapinbrightgreen.com/reports/
14-patterns/#rediscovering-the-intuitively-obvious

27

day twenty-seven: season

Poem Reading

Under Heaven

Find a market that sells a pomegranate
in early summer, and you find a place
that doesn't understand how appetite
has a season, how it takes the careful
cultivation of months for its many-chambered
heart to find fullness, a climate both steady and dry
to swell blossoms to galaxies wrapped in taut peel.
What true connoisseur hurries desire
or endures the pith, the grain-grind of seed
absent the anticipation of the small explosion
from the aril that purples the tongue?

—Anne M. Doe Overstreet,
   from *Delicate Machinery Suspended*

Prompt: Season

In "Under Heaven," the poet proposes that "appetite has a
season," and she reminds us that a pomegranate "takes the

careful cultivation of months" to come to fullness.

Using the specificity and passion of a "true connoisseur," write a poem about a fruit or vegetable that you love to eat in season. Describe its texture, taste, shape, and the climate and conditions it takes to produce it. If you're not sure about details, do a little research on your chosen food item. Feel free to compare it to the same food item eaten out of season.

A little side note about eating in season: often it takes less resources when you do so. For a fascinating discussion of the impacts of eating (or not eating) seasonally, check out Barbara Kingsolver's *Animal, Vegetable, Miracle: A Year of Food Life*.

28

day twenty-eight: care

Poem Reading

Contingency Plans

If we might figure out why the water waits
If we could reason our way out of this mess,

> Then we might never have reason to fear hurricanes.
> Then we could clean the cellar, closets, and the garage.

When we manage to stitch the night by all its stars,
When we finally think of names for everyone,

> Make sure to keep the knot from slipping out of sight.
> Make sure to tell each, *I love you, I love you, I love you.*

Suppose we try returning to times when God was God,
Suppose we decide our recycling ranks next to holiness,

> And I will never yield the floor for pleas for mercy.
> And I will try to keep the bins kempt and orderly.

—David K. Wheeler, from *Contingency Plans*

Prompt: Whose Care

Earth care can be fraught with competing philosophies and concerns: *we are responsible for nothing* (let the cosmos or the Divine take care of it) versus *we are responsible for everything* (it's up to us to "keep the bins kempt and orderly").

The truth of the matter probably lands somewhere in between. After all, we are part of nature and, like any other species, we are both acted upon by our environment and we act upon it. Similarly, if we are religious, there is some divine being or force that we both *ask things of* and who/which *asks things of us*.

A complicating factor in all of this is that we can begin to feel "godly" or "better-than" others, regardless of which philosophy we adhere to more closely. This can easily have the effect of turning others away, rather than bringing them along. It can also cloud our vision as to how to move forward in the most compassionate and effective ways.

Write two short poems about some aspect of earth care— the first part where you are responsible for nothing and the second part where you are responsible for everything.

Now, cut your poems apart into lines or phrases and put all the parts in one pile. From these cutouts, make a new poem.

See "Cut It Out Poems" at tweetspeakpoetry.com for an example of how to assemble a poem from disparate lines.

"Cut It Out Poems" can be found here:
https://www.tweetspeakpoetry.com/2013/01/10/taking-poetry-to-work-a-few-good-tricks/

29

day twenty-nine: running

Poem Reading

As the Deer

We owe it to each other
to share what white tail already know.
When the pressure changes, they run
together, hooves clacking across asphalt,
then silent on the dewy lawns.

—Marcus Goodyear, from *Barbies at Communion*

Prompt: Running With

Our poem reading for today uses the metaphor of deer—they "run together" to deal with "the pressure changes."

Write a poem about what we humans can do now that the "pressure" has changed and we understand the global climate problems we face. Feel free to choose a specific challenge we could focus on, rather than the more general issue of climate "problems." (There are 80 ideas about specific focus areas that you can find at PoeticEarthMonth.com. See the link at the end of this prompt. Or, see page 82 in this book.)

How can we run together? Is there something we can learn from other species, like the deer, about how to handle certain climate breakdown challenges? Choose a species as small as the ants or as large as the elephants, to learn from. Do a little research about your species so you can include some interesting specific details in your poem.

The Top 80 Game Changers can be found here:

http://poeticearthmonth.com/top-80-game-changers-in-curbing-climate-change/

30

day thirty: making

Poem Reading

Neruda's Memoirs, excerpt

2

Neruda said the closest thing to poetry
is a loaf of bread
or a ceramic dish
or a piece of wood lovingly carved

So he poured his words
into the glass of another language
only some of the world speaks

He gave light to the mines of Coquimbo.
Now they glitter like dew on a silver fish.

He left the smell of fresh ink and crisp paper
at the broker's, who traded his wife's voice
for a rainbow of lightning.

He melted the snow on broad-sided mountains
to water the dust on Santiago's tongue.

He found the blue of Chile's sky
in the bellies of volcanoes, its silence
in a guitar in Spain.

Neruda's the rush of roots
after a sudden breath

the warm tear on a face in love

the sound of adolescence missing a beat.

3

The sea could rise above him;
the wind make a sail of him.

He was too young
for the blackness of his dress,
the year climbing to its close.

But when asked,
the poet shaped in the man
remembers a ride on an empty road
and the color of rain in his childhood
and the look of a long-necked swan
that would not sing when it died
heavy in his arms one undone afternoon.

—Maureen E. Doallas, from *Neruda's Memoirs*

Prompt: The Making of a Poet & A Poet's Makings

Whenever we face challenges, we have the privilege of framing them in words—words that express our hopes, our losses, our dreams; words that transform our personal vision or the world's. These words can become a source of sustenance and discovery, for the sometimes long work of bringing to birth necessary change.

Earth care, which is also about self and other care, presents you just such a challenge. And, as a poet (or a poet in the making!), you have the chance to, like our poem reading says today, "pour your words into the glass"—to give light, to water, to find, to remember, to hold.

Write a poem about the making of you as a poet who cares about earth, self, and others. Describe what you have experienced over the past 29 days or what you hope to do in the future, as a poet who will sometimes write about earth, self, and other care. You might also want to include something about your style and themes—the poems you've made or hope to make.

What long-ago or recent memories of the natural world shape you for this vocation and shape the poems you've made or hope to make? You might like to include them.

Consider writing about yourself in the third person ("he," "she," or "they"), as if someone else is telling the story of *you as a poet* or the story of *your poetry*. You can then try writing in the second person ("you") and in the first person ("I"), to create a trinity of poems about the makings of you as a poet (and the things you've made as a poet).

What do you feel either too young or too old for, in embarking on this journey? Feel free to use phrases from the "Neruda's Memoir" excerpt to shape your trinity of poems.

See a helpful explanation of second person writing, here:
https://www.tweetspeakpoetry.com/2018/12/05/read-like-a-writer-second-person-narrative-voice-in-claudia-rankines-citizen-an-american-lyric/

after the challenge

earth to poetry: answering

Now that Earth has spoken *in, to,* and *through* your poetry, how can you answer with your heart and life? Who can you potentially team up with, to share this incredible, fascinating, enriching journey? What new adventures in living beautifully and cleanly can you seek—daily, monthly, yearly?

Take some time to journal your thoughts on these questions. Choose an area where you might like to begin learning and growing, and see if a friend, co-worker, or family member might like to join you. You might find it interesting to choose one of the highest impact areas by exploring the Top 80 Game Changers identified in *Drawdown: The Most Comprehensive Plan Ever Proposed to Reverse Global Warming*. You might also find it helpful to join EcoChallenge for their semi-yearly challenges that take a gaming approach to learning and doing.

All in all, the best journeys are filled with discovery and joy. I wish you both.

the top 80 game changers
shared in drawdown

Consider taking some steps on one of the Top 10 out of these Top 80 game changers for curbing climate breakdown, identified by the Drawdown Coalition.

The coalition is a group of top scholars, scientists, entrepreneurs, and advocates from across the globe that's mapping, measuring, modeling, and communicating about the best solutions to reverse climate breakdown, with the goal of reaching drawdown.

Drawdown is the moment when the concentration of greenhouse gases in the Earth's atmosphere starts to decline each year.

If you notice in the solutions list that follows, two of the biggest sectors are Electricity and Food.

Turning a light off. Unplugging a toaster. That's not so hard. And eating a vegetable-based meal or preventing food waste is completely within anyone's reach. Other solutions might take a little more doing. But, hey, why not get together with friends, family, or co-workers and explore some of those, too?

Solutions are ranked by the total amount of greenhouse gases they either avoid or remove from the atmosphere. Some of the proposed solutions, like nuclear and waste-to-energy, are considered "regrets" solutions or just transitional solutions, because they can include negative impacts.

Rank	Solution	Sector
1	Refrigerant Management	Materials
2	Wind Turbines (Onshore)	Electricity Generation
3	Reduced Food Waste	Food
4	Plant-Rich Diet	Food
5	Tropical Forests	Land Use
6	Educating Girls	Women and Girls
7	Family Planning	Women and Girls
8	Solar Farms	Electricity Generation
9	Silvopasture	Food
10	Rooftop Solar	Electricity Generation
11	Regenerative Agriculture	Food
12	Temperate Forests	Land Use
13	Peatlands	Land Use
14	Tropical Staple Trees	Food
15	Afforestation	Land Use
16	Conservation Agriculture	Food
17	Tree Intercropping	Food
18	Geothermal	Electricity Generation
19	Managed Grazing	Food
20	Nuclear	Electricity Generation
21	Clean Cookstoves	Food
22	Wind Turbines (Offshore)	Electricity Generation
23	Farmland Restoration	Food
24	Improved Rice Cultivation	Food
25	Concentrated Solar	Electricity Generation
26	Electric Vehicles	Transport
27	District Heating	Buildings and Cities
28	Multistrata Agroforestry	Food

29	Wave and Tidal	Electricity Generation
30	Methane Digesters (Large)	Electricity Generation
31	Insulation	Buildings and Cities
32	Ships	Transport
33	LED Lighting (Household)	Buildings and Cities
34	Biomass	Electricity Generation
35	Bamboo	Land Use
36	Alternative Cement	Materials
37	Mass Transit	Transport
38	Forest Protection	Land Use
39	Indigenous Peoples' Land Management	Land Use
40	Trucks	Transport
41	Solar Water	Electricity Generation
42	Heat Pumps	Buildings and Cities
43	Airplanes	Transport
44	LED Lighting (Commercial)	Buildings and Cities
45	Building Automation	Buildings and Cities
46	Water Saving – Home	Materials
47	Bioplastic	Materials
48	In-Stream Hydro	Electricity Generation
49	Cars	Transport
50	Cogeneration	Electricity Generation
51	Perennial Biomass	Land Use
52	Coastal Wetlands	Land Use
53	System of Rice Intensification	Food
54	Walkable Cities	Buildings and Cities
55	Household Recycling	Materials

56	Industrial Recycling	Materials
57	Smart Thermostats	Buildings and Cities
58	Landfill Methane	Buildings and Cities
59	Bike Infrastructure	Buildings and Cities
60	Composting	Food
61	Smart Glass	Buildings and Cities
62	Women Smallholders	Women and Girls
63	Telepresence	Transport
64	Methane Digesters (Small)	Electricity Generation
65	Nutrient Management	Food
66	High-speed Rail	Transport
67	Farmland Irrigation	Food
68	Waste-to-Energy	Electricity Generation
69	Electric Bikes	Transport
70	Recycled Paper	Materials
71	Water Distribution	Buildings and Cities
72	Biochar	Food
73	Green Roofs	Buildings and Cities
74	Trains	Transport
75	Ridesharing	Transport
76	Micro Wind	Electricity Generation
77	Energy Storage (Distributed)	Electricity Generation
(77)	Energy Storage (Utilities)	Electricity Generation
(77)	Grid Flexibility	Electricity Generation
78	Microgrids	Electricity Generation
79	Net Zero Buildings	Buildings and Cities
80	Retrofitting	Buildings and Cities

about ecochallenge

EcoChallenge.org is designed to help people develop new ways of being and doing when it comes to areas like transportation, food, materials, energy use, and other lifestyle choices that affect emissions and the health of the earth and ourselves.

Several times a year, the organization offers 21-day challenges where you can *learn* and *do,* for a healthier you and a healthier world. During the challenges, you can create or join a team, choose your positive actions, earn points, share your story, and see how your collective impact adds up.

On their website, EcoChallenge notes, "Common wisdom says it takes at least three weeks to change a habit: if you can stick with a new behavior for 21 days in a row, you're a lot more likely to keep it up forever. EcoChallengers share their progress and earn points for taking action. The combination of collective inspiration, camaraderie and friendly competition makes change a little easier—and a lot more fun."

celebrate poetic earth month

T. S. Poetry Press invites you to *stay poetic* on your earth-, self-, and other-care journey, by celebrating Poetic Earth Month every April.

Also, throughout the year, you're invited to visit poetic earthmonth.com for inspiration, ideas, and resources, or come up with your own cool poetic ideas and tell TS about them. Use the hashtag #poeticearthmonth on social media and add an @tspoetry, so TS can see and share how you celebrate in April or stay poetic throughout the year!

bonus essay

it's not about what you can't do—
it's about love

These days I am often talking about climate breakdown with my best friends. (Lucky them. :)

Some people say that's the number one thing you can do if you want to make a difference. Talk to your friends and family about the subject. (Humorous climate scientist Katharine Hayhoe—yes there *is* such a thing as a humorous climate scientist—has a great TED talk about the power of honest-to-goodness talk that connects instead of divides.)

Just the other day, one of my friends remarked, "That's not my wheelhouse," when we were talking about something or other to do with climate. I was intrigued. How could this be? I look to this person for so many ideas about so many things. Surely there was something they could identify with, something they could speak to or do.

I mused on this all evening. Then it hit me. The climate issue is too often framed as what we *can't* do. It feels like it's way beyond us—whether that's because we don't know how to conceive of gigatons of CO_2 or whether that's because there doesn't seem to be one go-to handbook on how to change all this. (Well, maybe now there is [*Drawdown*], but even so, it's not always clear what you and I can do right now. Today.)

I think this whole *can't do/can do* thing should be less about science-mindedness and strategy, in the end. And more about love. But hardly anyone talks about it this way.

My favorite woman ever, to illustrate the love point, is Wangari Maathai. She grew up in Africa, with trees as her companions and comforters. Every day, says children's-book-writer Franck Prévot, Wangari fetched water "at the foot of the big mugumo, the generous fig tree."

I love figs when they are in Fig Newtons. I don't know if Wangari ever tasted a Fig Newton. She did come to the States to study for a while, which was a great and unusual opportunity for an African woman of her generation. But I don't know if her time here afforded her the chance for a Fig Newton. Regardless, in the shade of the big fig tree, she learned from her mother that "a tree is worth more than its wood." Says Prévot, this was an expression Wangari never forgot.

That is true that a tree is worth more than its wood, but it is also true that some of us love wood best of all. And some of us love Fig Newtons, too. More than trees. And I think that is perfectly okay. Because if you truly love wood and Fig Newtons, then what you can do is help to make sure there will be plenty of it (and them) in the world going forward. Just like Wangari, you can plant a tree. (Or you can let one spring up.) It can be a cherry tree or an oak. It can be a fig tree, if you have the right climate for that.

Personally, I have planted nine trees on my postage-stamp-sized property, and I have let a few more spring up where they're "not supposed to be." They cool the space around my house. They provide color and texture. A few give me fruits. And I am doing my small part towards making the air breathable into the future, for you and me. The little lemon tree inside the house? That's just for fun. This year we had a 100%

increase in production. Two lemons!—over the one we got in the past. I'm a regular citrus green thumb.

Which brings me back to Wangari Maathai.

The truth about *me* is that I am a better writer than I am a gardener. I am a better let-it-spring-up planter than I am an arborist. But Wangari? She was good at tree planting. It was her wheelhouse. And because of that, with the help of countless women she empowered and trained, she successfully planted millions (and millions and millions) of trees in Kenya, reforesting great swathes of the country.

What does this mean for you and me—if we aren't exactly like Wangari? I don't want to talk to us about what we *can't* do. I want to talk to us about what we *can* do. And what we *can* do, if we are going to do it with some level of aplomb and commitment, needs to stem from love, like it did for Wangari. I'm not saying that means we won't have to learn something. Love certainly draws us on, making learning feel more like discovery and making work feel worth it, even when that work doesn't exactly feel like play.

In Prévot's *Wangari Maathai: The Woman Who Planted Millions of Tree*s, you can read about where Wangari's love for trees took her in life. There were some hard roads. But there was joy, too. And you and I can breathe just a bit easier, thanks to her foresight in planting trees.

Speaking of wise and wonderful foresight, there's an effort underway, to plant more than a trillion trees, to make it so you and I and our kids and their kids can not only breathe, but can also smooth their hands across a cherry table or enjoy a Fig Newton under a maple tree—or maybe under a cedar, depending on the region where we grow and change and lean back to our past and continue to dream of the world to be. What can *you* do? Start by considering what you love.

—*L.L. Barkat, who loves trees, 2019*

endnotes

Epigraph

p. i   "Poetry is a life-cherishing...": Mary Oliver, *A Poetry Handbook: A Prose Guide to Understanding and Writing Poetry* (Boston, Mariner Books, an imprint of Houghton Mifflin Harcourt, 1994), p. 122.

p. i   "I think there's a kind of desperate...": W. S. Merwin, W. S. Merwin biography. (Accessed online at www.poetryfoundation.org.) https://www.poetry foundation.org/poets/w-s-merwin

Day 2

p. 16   "Today I am startled again...": Jennifer A. Francis, "The Arctic Is Breaking Climate Records, Altering Weather Worldwide" in *Scientific American* (New York, New York, Vol 319, No 4, April 2018).

p. 17   The Top 80 Game Changers: Paul Hawken and Tom Steyer. *Drawdown: The Most Comprehensive Plan Ever Proposed to Reverse Global Warming* (New York, Penguin Books, an imprint of Penguin Random House LLC, 2017), p. 222-223.

Day 3

p. 19   "There is always something left...": Lorraine Hansberry, *A Raisin in the Sun* (Vintage, a division of Random House, 2004), p. 145.

Day 26

p. 69   "Biophilic design can reduce stress...": Terrapin Bright Green, in "14 Patterns of Biophilic Design" (Accessed online at www.terrapinbrightgreen.com.) https://www.terrapinbrightgreen.com/reports/14-patterns/#rediscovering-the-intuitively-obvious

Bonus Essay

p. 88   The humorous TED talk by Katharine Hayhoe is "The Most Important Thing You Can Do to Fight Climate Change: Talk About It." It can be accessed at https://www.ted.com/talks/katharine_hayhoe_the_most_important_thing_you_can_do_to_fight_climate_change_talk_about_it#t-1019746

p. 90   *Wangari Maathai: The Woman Who Planted Millions of Trees*, by Franck Prévot (Author), Aurélia Fronty (Illustrator) (Watertown, MA, Charlesbridge, 2017).

p. 90   The Trillion Tree campaign can be viewed at https://www.trilliontreecampaign.org/

acknowledgements

To the Tweetspeak Poetry* team, thank you for the work you do every day, that brings thoughtfulness, beauty, and fun to the world. Thank you, also, for always being ready to grow, as you were when I brought the idea of Poetic Earth Month to you. Without your support and dedication, this kind of new direction would not be possible—nor would it be possible for me to "disappear" for two full months to develop projects like the 30-Days, 30-Poems Challenge that first ran at poeticearth month.com and which has now become part of this book.

To our Tweetspeak patrons who have given extra and to others who have also given towards making Poetic Earth Month happen, thank you, especially:

Bill and Linda Thorla
Will Willingham
John and Megan Willome
Glynn Young
Rhonda Danette Owen
S. Savage
Laura Lynn Brown
Maureen E. Doallas
Sandra Heska King
Dheepa R. Maturi
Richard Maxson
Michelle Ortega
Bethany Rohde
Susan Goldsmith Wooldridge
Debra Hale-Shelton

To all of our Tweetspeak patrons, who help keep what I fondly call "The Mother Ship" afloat, thank you. Your partnership works towards keeping Tweetspeak viable in the world and makes it possible for us to grow in additional ways, as we have with the development of Poetic Earth Month and its associated projects like this book; to those patrons among you who give at the highest levels, a special thanks:

Michelle Ortega
Rhonda Danette Owen
Sandra Heska King
Mary L Van Denend
Will Willingham
John and Megan Willome
Glynn Young

*Tweetspeak Poetry, at tweetspeakpoetry.com, is a literary and writing site published by T. S. Poetry Press.

permissions & credits

Poems by L.L. Barkat, Maureen E. Doallas, Marcus Goodyear,
Dave Malone, Zoya Marchenko, Ruth Mowry, Anne M. Doe
Overstreet, and David K. Wheeler are used with permission
from T. S. Poetry Press. They appeared in the following books:

Barbies at Communion, by Marcus Goodyear, 2010.

Contingency Plans: Poems, by David K. Wheeler, 2010.

Delicate Machinery Suspended: Poems, by Anne M. Doe Overstreet,
2011.

God in the Yard: Spiritual Practice for the Rest of Us, by L.L. Barkat,
2010.

*How to Read a Poem: Based on the Billy Collins Poem "Introduction to
Poetry,"* by Tania Runyan, 2014.

Love, Etc.: Poems of Love, Laughter, Longing & Loss, by L.L.
Barkat, 2014.

Neruda's Memoirs: Poems, by Maureen E. Doallas, 2011.

O: Love Poems From the Ozarks, by Dave Malone, 2015.

Sun Shine Down: A Memoir, by Gillian Marchenko, 2013.

The Joy of Poetry: How to Keep, Save & Make Your Life With Poems, by Megan Willome, 2016.

The Novelist: A Novella, by L.L. Barkat, 2012.

"Back" first appeared in *Every Day Poems* and is used with permission of Will Willingham.

"Come night…" is from *Romeo & Juliet*, by William Shakespeare and is in the public domain.

"It's Not About What You Can't Do—It's About Love" was first published at PoeticEarthMonth.com on March 13, 2019. Used with permission of L.L. Barkat.

Newtons (colloquially called "Fig Newtons") are a cookie trademarked and produced by Nabisco. Nabisco is a subsidiary of Illinois-based Mondelez International.

"Night" by Sara Teasdale is in the public domain.

"The Lake Isle of Innisfree" by William Butler Yeats is in the public domain.

"Today By the Creek I Pretend I'm the Dalai Lama" first appeared in *Every Day Poems* and is used with permission of Susan Goldsmith Wooldridge.

"Turning" is used with permission of Laurie Klein.

also from t. s. poetry press

How to Read a Poem: Based on the Billy Collins Poem "Introduction to Poetry," by Tania Runyan

Runyan's book reads like a playful love letter—a creative intercession on poetry's behalf—to the hearts of a new generation, those on whom so much, like the future of the art, depends.

—Brad Davis, Poet, teacher, and counselor at Pomfret School

On Being a Writer: 12 Simple Habits for a Writing Life that Lasts, by Ann Kroeker and Charity Singleton Craig

A genial marriage of practice and theory. For writers new and seasoned. This book is a winner.

—Philip Gulley, author of *Front Porch Tales*

The Teacher Diaries: Romeo & Juliet, by Callie Feyen

Feyen weaves memoir with educational strategy, presenting it with humor, empathy, and an urgency that helps us see the relevance of Shakespeare's most famous tragedy.

—Cara Gabriel, Assistant Professor, Department of Performing Arts at American University